Meteorological Predictions and Climatological Factors

Enhanced Predictive Models

One of the foremost reasons why the 2024 hurricane season might be significant is due to advancements in meteorological science and predictive models.

Recent years have seen considerable improvements in the accuracy of weather forecasting, thanks to enhanced computational models, satellite technology, and data assimilation techniques.

These advancements allow meteorologists to predict the formation, trajectory, and intensity of hurricanes with greater precision, thus providing better preparedness measures for potentially affected areas.

Significant 2024 Hurricane Season Attention

The 2024 hurricane season has already garnered significant attention due to a combination of climatic factors, emerging weather patterns, and the potential socio-economic impact it might have.

Analyzing the significance of this hurricane season involves delving into various aspects such as meteorological predictions, historical comparisons, the role of climate change, and the implications for affected regions.

This extensive exploration aims to provide a comprehensive understanding of why the 2024 hurricane season could be particularly noteworthy.

El Niño and La Niña Oscillations

The interplay between El Niño and La Niña conditions in the Pacific Ocean significantly influences hurricane activity in the Atlantic.

In 2024, predictions indicate a potential transition from a La Niña phase to an El Niño phase.

La Niña conditions, characterized by cooler-than-average sea surface temperatures in the central and eastern Pacific, typically enhance Atlantic hurricane activity due to reduced vertical wind shear.

Conversely, El Niño conditions suppress hurricane activity by increasing wind shear across the tropical Atlantic.

The transitional phase between these oscillations can create unpredictable and potentially volatile weather patterns, leading to a heightened possibility of severe hurricanes.

Sea Surface Temperatures

Another crucial factor is the current trend of rising sea surface temperatures (SSTs).

Warmer ocean waters serve as the primary energy source for hurricanes, facilitating their formation and intensification.

In 2024, SSTs in the Atlantic are expected to remain above average, providing favorable conditions for the development of stronger and more frequent hurricanes.

This trend is consistent with broader global warming patterns, which have been linked to increased hurricane activity and intensity.

Atmospheric Conditions

In addition to SSTs, atmospheric conditions such as humidity levels, atmospheric pressure, and wind patterns play pivotal roles in hurricane formation.

The presence of a favorable Madden-Julian Oscillation (MJO) phase, characterized by enhanced convection and precipitation in the tropical regions, can also contribute to an active hurricane season.

The alignment of these factors in 2024 suggests the potential for a significant and impactful hurricane season.

Historical Comparisons and Trends

Recent Active Seasons

A historical perspective further underscores the potential significance of the 2024 hurricane season.

In recent years, there has been a noticeable increase in the frequency and intensity of hurricanes.

The 2020 season, for instance, was one of the most active on record, with 30 named storms, 13 hurricanes, and six major hurricanes.

The 2021 season, while not as prolific, still produced several notable hurricanes, including Ida, which caused widespread devastation.

These trends point towards an ongoing pattern of heightened hurricane activity, raising concerns about the upcoming 2024 season.

Long-term Climate Trends

Long-term climate data reveals an upward trend in the intensity and destructiveness of hurricanes.

This trend is attributed to various factors, including global warming, which leads to higher SSTs and increased atmospheric moisture.

As a result, hurricanes are becoming more powerful, with higher wind speeds, greater rainfall, and longer durations.

The 2024 season, set against this backdrop of escalating hurricane activity, may follow this trend and produce several high-impact storms.

The Role of Climate Change

Global Warming and its Impact

Climate change is a critical factor that cannot be ignored when discussing the significance of the 2024 hurricane season.

Global warming, driven by the increased concentration of greenhouse gases in the atmosphere, has led to a rise in global temperatures.

This warming effect is particularly pronounced in the oceans, where higher SSTs provide more energy for hurricane formation and intensification.

The result is a tendency for hurricanes to become more severe, with greater potential for catastrophic damage.

Increased Moisture Content

One of the direct consequences of global warming is the increased moisture content in the atmosphere.

Warmer air holds more moisture, leading to heavier rainfall during hurricanes.

This increased rainfall can cause severe flooding, compounding the damage caused by high winds and storm surges.

The 2024 hurricane season, with its elevated SSTs and warmer atmospheric conditions, is likely to produce storms with significant rainfall, raising concerns about flooding in vulnerable areas.

Rising Sea Levels

Rising sea levels, another consequence of climate change, exacerbate the impact of hurricanes.

Higher sea levels mean that storm surges, which are already a dangerous component of hurricanes, can penetrate further inland, affecting more areas and increasing the potential for damage.

Coastal communities, which are often the first to bear the brunt of hurricanes, are particularly at risk.

The 2024 season, with the ongoing rise in sea levels, could see more extensive and severe storm surge impacts.

Socio-economic Implications

Vulnerable Populations

The significance of the 2024 hurricane season extends beyond meteorological and climatological factors to include socio-economic implications.

Vulnerable populations, particularly those in low-lying coastal areas, are at heightened risk.

These communities often lack the resources and infrastructure to effectively prepare for and respond to hurricanes, making them more susceptible to the devastating effects of these storms.

The potential for a severe hurricane season in 2024 raises concerns about the ability of these populations to withstand and recover from such events.

Economic Impact

Hurricanes have substantial economic impacts, causing billions of dollars in damage to infrastructure, homes, businesses, and agricultural sectors.

The 2024 hurricane season, with the potential for high-intensity storms, could lead to significant economic losses.

The cost of recovery and rebuilding can strain local and national economies, diverting resources from other critical needs and impeding long-term development efforts.

The economic repercussions of a severe hurricane season can also extend globally, affecting markets, supply chains, and trade.

Insurance and Reinsurance Markets

The insurance and reinsurance markets are particularly sensitive to hurricane activity.

High-frequency and high-intensity hurricanes can lead to substantial insurance claims, affecting the profitability and stability of insurance companies.

In response, insurance premiums may increase, and coverage options may become more limited, particularly in high-risk areas.

The 2024 hurricane season, if it proves to be as significant as anticipated, could have lasting effects on the insurance industry, influencing policy decisions and coverage availability.

Public Health and Safety

Public health and safety are paramount concerns during hurricane seasons.

Hurricanes can lead to direct physical harm through high winds, flying debris, and flooding.

Additionally, they can disrupt access to essential services such as healthcare, clean water, and sanitation.

The 2024 hurricane season, with its potential for severe storms, poses significant risks to public health and safety.

Effective disaster preparedness and response measures are crucial to mitigating these risks and protecting vulnerable populations.

Preparedness and Response

Emergency Management

The potential significance of the 2024 hurricane season underscores the importance of robust emergency management strategies.

Governments, emergency services, and communities must be well-prepared to respond to hurricane threats.

This includes having effective evacuation plans, adequate shelters, and efficient communication systems in place.

The ability to quickly mobilize resources and provide assistance to affected areas can significantly reduce the impact of hurricanes and save lives.

Infrastructure Resilience

Building resilient infrastructure is another critical aspect of hurricane preparedness.

This involves constructing buildings and infrastructure that can withstand high winds and flooding, as well as retrofitting existing structures to improve their resilience.

In addition, investing in natural defenses such as wetlands and mangroves can provide additional protection against storm surges and flooding.

The 2024 hurricane season, with its potential for severe storms, highlights the need for continued investment in infrastructure resilience.

Community Awareness and Education

Community awareness and education play vital roles in hurricane preparedness.

Ensuring that residents are informed about hurricane risks and know how to respond can significantly enhance community resilience.

Public education campaigns, emergency drills, and community engagement initiatives can help residents prepare for hurricanes and respond effectively when they occur.

The significance of the 2024 hurricane season emphasizes the need for ongoing efforts to raise awareness and educate communities about hurricane preparedness.

Environmental Impact

Ecosystem Disruption

Hurricanes can have profound impacts on ecosystems, causing damage to forests, wetlands, coral reefs, and other natural habitats.

The high winds, heavy rainfall, and storm surges associated with hurricanes can lead to erosion, habitat loss, and changes in species composition.

The 2024 hurricane season, with its potential for severe storms, could result in significant environmental disruption.

Protecting and restoring natural ecosystems is essential for maintaining biodiversity and mitigating the impacts of hurricanes.

Climate Feedback Loops

Hurricanes also play a role in climate feedback loops, influencing global weather patterns and climate systems.

For example, hurricanes can transfer heat from the ocean to the atmosphere, affecting atmospheric circulation and weather patterns.

Additionally, the destruction of forests and other carbon sinks during hurricanes can release stored carbon dioxide, contributing to global warming.

The potential for an active and intense hurricane season in 2024 highlights the complex interplay between hurricanes and climate systems.

Global Perspective

International Aid and Cooperation

The significance of the 2024 hurricane season extends beyond national borders, highlighting the importance of international aid and cooperation.

Hurricanes can affect multiple countries, particularly in regions such as the Caribbean and Central America.

International organizations, governments, and non-governmental organizations (NGOs) play crucial roles in providing aid and support to affected countries.

Effective coordination and collaboration are essential for delivering timely assistance and supporting recovery efforts.

Climate Change Mitigation and Adaptation

The potential for a significant 2024 hurricane season underscores the urgent need for global efforts to mitigate and adapt to climate change.

Reducing greenhouse gas emissions, transitioning to renewable energy sources, and implementing sustainable land-use practices are critical for addressing the root causes of climate change.

Additionally, investing in climate adaptation measures, such as building resilient infrastructure and protecting natural ecosystems, is essential for enhancing resilience to hurricanes and other climate-related impacts.

Conclusion

The 2024 hurricane season holds the potential to be particularly significant due to a combination of meteorological, climatological, socio-economic, and environmental factors.

Advances in predictive models, the influence of El Niño and La Niña oscillations, rising sea surface temperatures, and changing atmospheric conditions all contribute to the potential for an active and severe hurricane season.

Historical trends and long-term climate data further underscore the likelihood of heightened hurricane activity.

Climate change plays a central role in shaping the intensity and frequency of hurricanes, with global warming leading to warmer ocean waters, increased atmospheric moisture, and rising sea levels.

The socio-economic implications of a significant hurricane season are substantial, affecting vulnerable populations, economies, insurance markets, and public health and safety.

Preparedness and response measures, including robust emergency management strategies, resilient infrastructure, and community awareness and education, are crucial for mitigating the impact of hurricanes.

The environmental impact of hurricanes, including ecosystem disruption and climate feedback loops, highlights the need for protecting natural habitats and addressing the complex interactions between hurricanes and climate systems.

From a global perspective, international aid and cooperation, as well as efforts to mitigate and adapt to climate change, are essential for addressing the challenges posed by hurricanes.

The potential significance of the 2024 hurricane season serves as a stark reminder of the interconnectedness of our world and the need for collective action to build resilience and protect our planet.

In summary, the 2024 hurricane season could be significant due to a confluence of factors that create the conditions for potentially severe and impactful storms.

Understanding these factors and their implications is essential for preparing for and responding to the challenges that this hurricane season may bring.

Please use the next few pages for your notes and debates.

www.ingramcontent.com/pod-product-compliance
Lightning Source LLC
Chambersburg PA
CBHW071959210526
45479CB00003B/992